William Blizard

A Lecture, on the Situation of the Large Blood-Vessels of the Extremities

And the methods of making effectual pressure on the arteries, in cases of

dangerous effusions of blood from wounds

William Blizard

A Lecture, on the Situation of the Large Blood-Vessels of the Extremities
*And the methods of making effectual pressure on the arteries, in cases of
dangerous effusions of blood from wounds*

ISBN/EAN: 9783337393250

Printed in Europe, USA, Canada, Australia, Japan

Cover: Foto ©berggeist007 / pixelio.de

More available books at **www.hansebooks.com**

A

L E C T U R E,

On the Situation of the large Blood-Veffels of the Extremities; and the Methods of making effectual Preffure on the Arteries, in Cafes of dangerous Effufions of Blood from Wounds:

DELIVERED TO THE SCHOLARS OF

THE LATE MARITIME SCHOOL AT CHELSEA;

And firft printed for their Ufe.

THIRD EDITION.

TO WHICH IS NOW ADDED,

A brief Explanation of the Nature of *Wounds*,

More particularly thofe received from FIRE-ARMS.

BY WILLIAM BLIZARD, F.R.S.

Prodeffe quàm confpici.

LONDON:

Printed by H. L. GALABIN, Ingram-Court;
and publifhed by C. DILLY, Poultry.

1798.

IT is the duty of every man to be ready to meet the enemies of peace, order, and happiness; but, while preparing to affume all the noble Britifh charaƈter, through which, under Divine Providence, our matchlefs conftitution and unparalleled bleffings have been acquired and continued, let us not be unmindful of thofe means of prefervation, in fituations of danger, that Science and Art direƈt, of which the following are at all times proper to be underftood, and efpecially at the prefent moment.

Devonfhire-Square,
April 30, 1798.

PREFACE.

PREFACE.

THE INTRODUCTION prefixed to thefe pages, when firft printed for the ufe of the fcholars of the late MARITIME SCHOOL at Chelfea,* explains their original defign. A paffage in Captain DRINKWATER's Account of the Siege of Gibraltar, expreffes the probable utility of fuch a publication. —— "Sep- "tember, 1781. The 30th, a foldier of

* An inftitution intended for the maintenance and nautical inftruction of the fons of thofe naval officers who had bravely fallen in the fervice of their country, without a provifion for the fupport and education of their children. The failure of this undertaking is to be lamented as a national misfortune. May public fpirit foon revive the humane and patriotic defign; to remain a monument of regard for thofe objects that ought to be held moft dear by Englifhmen!

" the

" the 72d. loft his legs by a fhot from
" Fort *Barbara*. He bore amputation
" with prodigious firmnefs ; but died, foon
" after, through the lofs of blood pre-
" vioufly to his being brought to the hof-
" pital. This fact being reprefented to
" the governor, the fergeants of the dif-
" ferent regiments were ordered to attend
" the hofpital, to be taught by the fur-
" geons how to apply the TOURNIQUET ;
" which was afterwards productive of very
" beneficial confequences. Tourniquets
" were alfo diftributed to the different
" guards, to be at hand in cafe of ne-
" ceffity."*

Were the knowledge of the fituation of
the blood-veffels of the extremities, fo far
as is neceffary for checking dangerous effufions of blood, and the ufe of the tourniquet, more general ; not confined to the
navy and army, but extended to col-
leges and fchools, particularly military and

* Vide Drinkwater's Hiftory of the Siege of Gibraltar, p. 190.

nautical

nautical academies, manufactories, hofpi-
tals of every defcription, prifons, planta-
tions, fire-offices, the clergymen of pa-
rifhes in which no furgeons are refident,
commanders of merchantmen, miners, &c.
it could not fail of proving highly beneficial to mankind.

The late Sir BARNARD TURNER would
have bled to death, on the fpot of his accident that terminated fatally, had not compreffion been inftantly made on the artery
of the wounded limb. Laft winter, a
poor man, in Cornhill, actually bled to
death, from a ruptured veffel in his leg, for
want of the timely application of a tourniquet. —— But the experience of moft
perfons could afford inftances of danger
or death through defect of this knowledge.

When a fellow-creature is reftored from
a ftate of apparent extinction of life by
drowning, rewards are affigned to thofe
who exerted themfelves in the recovery.
The knowledge of the means proper to

be

be employed on fuch an alarming occafion is alfo, very humanely, generally
propagated. Surely, then, if men be in
earneft in their endeavours for the prefervation of human life, they will admit
the importance of the information here
recommended ; fince there is no doubt
that many have fallen facrifices to ignorance of the means of reftraining HÆ-
MORRHAGE.

The familiar form of the Lecture is
retained, as the beft for general information.

July 30, 1786.

INTRO-

INTRODUCTION.

FROM reflection on my duty, as SUR-GEON to the MARITIME SCHOOL, and a sincere regard for the objects of my care, I proposed to teach them the situation of the large blood-veffels of the extremities, and the application of the TOURNIQUET. This I attempted, in the plaineft manner in my power, in the way of LECTURE, as the moft familiar and effectual method of impreffing truths on juvenile minds: and it was pleafing to obferve the ATTENTION and FEELING expreffed by my young auditors.

From a defire to promote the great caufe of the naval intereft of my country, in that effential concern, THE PRE-SERVATION OF THE LIVES OF SEA-MEN, I have now endeavoured to render my Lecture an ufeful OFFERING to thefe young warriors.

In

In the navy and army, cafes continually occur, in which the information it contains is abfolutely neceffary for the prefervation of exiftence : but there can hardly be a fituation of life, in which, at fome period, the knowledge might not prove of equal importance ; and it cannot fail of adding to confidence and courage in the moment of danger.

But knowledge of this kind may be productive of fome degree of good, though never *practically* required; for, SCIENCE ever tends to improve the heart, and raife the mind to contemplate the power, wifdom, and goodnefs, of HIM THAT MADE US!

No profeffional fame can be acquired from explaining facts known to every ftudent in furgery. This little work muft, therefore, be confidered as a tribute to HUMANITY, offered from a fenfe of duty.

July 15, 1783.

A LEC-

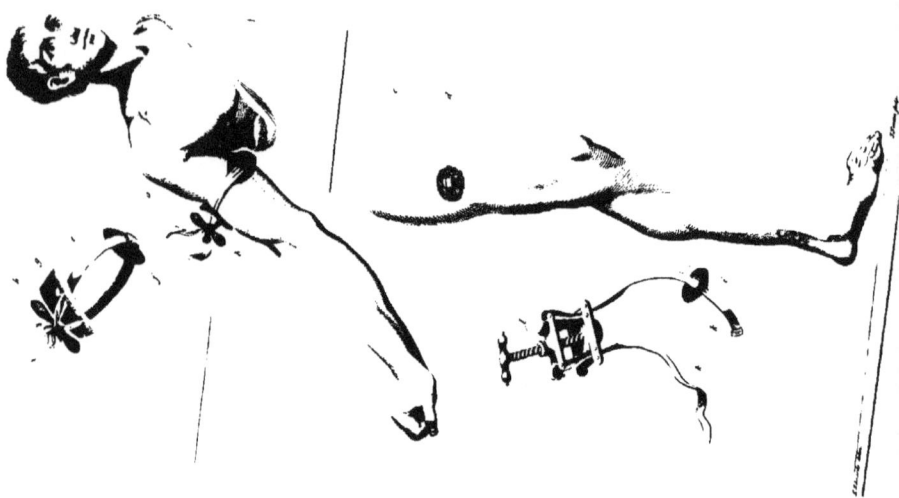

LECTURE, &c.

YOUNG GENTLEMEN,

AS one of the guardians of your health and lives, I requeſt your attention, while I point out what may conduce to the preſervation of theſe bleſſings when you are launched into the world, as well as during your reſidence in this ſeminary of naval ſcience.

You are here educated to a profeſſion of great honour, becauſe of high utility. It is the ſecurity of our country, our religion and laws, our commerce and riches.

riches. The SEAMAN, then, according to his rank and merit, has a claim to the refpect and care of his countrymen.

You are ambitious to become SEAMEN, are ready to join the veteran band, to go forth to fight the enemies of your country, and therefore merit the efteem and fervices of your fellow-citizens.

We are excited to attend to the welfare of the BRITISH SAILOR by another confideration. Trained up in the principles of true honour and bravery, hardy in the practice of them, and properly confidering his life as devoted to the fervice of his country, he is lefs mindful of bodily evils, and the means of averting them, than the more wary and delicate landfman. He has a title, then, in generofity, to that attention from others which a martial fpirit prevents him from fhewing to himfelf.

I am affured, gentlemen, that, in his majefty's fhips, you will have many oc-

caſions

cafions for the exercife of your judgement and fpirit refpecting the health
and lives of your men. You muft *re-
flect for them*; and, when they find that
you are truly zealous in all things for
their good, they will obey with ala-
crity, will bear you with fpirit through
danger, and prove themfelves worthy of
your generous regard. —— Thefe confiderations will, I truft, engage your attention to whatever promifes benefit to your
companions in war.

Every good and brave man would lay
down his life in the difcharge of his duty
to his king and country. But, when
fick or hurt, he is not to neglect the
means of relief which PROVIDENCE has
afforded. On the contrary, we are com-
manded, by divine authority, to preferve
our lives and thofe of our fellow-crea-
tures.

For the prefervation of the health and
lives of the officers and feamen of his
majefty's navy, there are appointed, by
government,

government, to each ſhip of war, a SUR-
GEON, and a certain number of MATES
according to the rate of the ſhip. During the time of action, the ſtation of
theſe officers is in .the COCK-PIT. From
their neceſſary confinement to this ſituation, evils of a very ſerious nature may
ſometimes happen ; for they cannot poſ-
ſibly render inſtantaneous aſſiſtance to thoſe
in a remote part of the veſſel, whoſe bleeding wounds may urgently require the aid
of ſurgery.

Some of the methods of chirurgical relief are very ſimple, though of the great-
eſt importance. Of this kind is the making an effectual temporary preſſure upon
a part, to prevent a fatal effuſion of blood,
in the caſe of wound, till means of perma-
nent benefit can be employed.

Men of true courage are not diſmayed
at the ſight of blood. In firm poſſeſſion
of themſelves, on all occaſions, they are
capable of exerciſing their judgement, and
employing the means with which they
are

are happily acquainted, either to their own benefit or that of others. It is proper, then, that they fhould have information of whatever is ufeful, and in their power to execute.

I cannot omit this opportunity, my young friends, of exhorting you to be EXAMPLES OF SOBRIETY as well as of the other VIRTUES. What advantage can flow from reafon or courage in a ftate of intoxication ? Many a brave fea-man has loft his life from having his mind clouded, by the effects of ftrong liquor, at the time of receiving a wound. — By TEMPERANCE the body is preferved free from various diforders, and the mind calm and firm, to direct under circumftances of accidents and on every trying occa-fion.

Induced by thefe confiderations, I pro-pofed to the good men who direct your education, to teach you the application of the inftrument, called TOURNIQUET, employed for ftopping the flow of blood from

from wounded veſſels. With their ſanc-
tion, I have the pleaſure of addreſſing you
on this ſubjeƈt, and moſt heartily wiſh the
inſtruƈtion may prove uſeful.

A circumſtance has occurred, ſince I
propoſed to meet you on this occaſion,
which has ſtrengthened my notions re-
ſpeƈting the utility of the intended expla-
nations; and will, I have no doubt, be ſa-
tisfaƈtory to your governors.

I requeſted the ſentiments of an intelli-
gent naval ſurgeon on the ſubjeƈt. This
was his anſwer:

" I can beſt expreſs my opinion by
" relating to you the praƈtice of an in-
" genious ſurgeon in the ſervice, and aſ-
" ſuring you that his and my ſentiments
" perfeƈtly coincide. —— Mr. ****, ſur-
" geon of the BARFLEUR, had obſerved,
" with great concern, the dreadful effeƈts
" of wounds that happened in time of
" aƈtion, from the ſeamen being entirely
" ignorant of the manner of applying
" the tourniquet, many inſtances having
" occurred

" occurred of men bleeding to death, par-
" ticularly in the tops, before affiftance
" could poffibly be rendered them. ——
" To prevent thefe evils, as much as
" was in his power, he provided every
" feaman, ftationed in the tops, with a
" tourniquet; and, on every opportu-
" nity, taught them the method of ap-
" plying it; fo that, in a fhort time,
" they became perfectly expert in its
" ufe."

The pious Pfalmift beautifully exclaims,
" I am fearfully and wonderfully made!"
It would, indeed, require the ftudy of a
long life to learn the little that has been
difcovered of INFINITE WISDOM in the
ftructure of the feveral parts of the human body, and of INFINITE GOODNESS
in the laws by which they perform their
functions to the maintenance of health and
life.

It is proper, however, that you fhould
have a general idea of the circulation of
the blood, in order to underftand the

practice that will be laid down, and
to enable you to adapt it to particular
cafes.

" In the BLOOD is the LIFE of man."
That is to fay, this fluid contains the
principles of nourifhment, and diftributes
them to every part of the body for its fupply and refrefhment ; like the water of the
great ocean, which conveys the riches and
good things of the world to every quarter
of the globe.

The HEART is the fource of this fluid.
It is feated in the breaft, a little to the left
fide ; nearly, however, in the centre of the
body. This organ is hollow, for containing the blood; and it has the power of
contracting, and ftrongly propelling its
contents. By this contraction of the heart,
the blood is pufhed forwards, with an exceedingly rapid current, to the remoteft
parts of the body; as the tide of the
fea influences and preffes on the waters
of rivers, obfervable here in the fwelling
Thames.

The

The veffels, or tubes, which proceed from the heart, to convey the blood to all the parts of the body, are called AR- TERIES. From the power with which the heart propels the blood through this fyftem of veffels, it happens, that, when- ever they are wounded, the blood flows rapidly and in jerks from the wounded part. They divide, to be diftributed to parts, from trunks, like the branches of a tree from its body; fo that, on pref- fing together the fides of any trunk, the flow of blood, into the branches beyond the compreffed part, is prevent- ed.

The veffels, which return the blood to the heart, are named VEINS. The blood in them receives but little of the impelling force of the heart, and, there- fore, moves not with a ftrong tide or current, but glides evenly and gently on, like the ebbing water; and, confequent- ly, wounds of thefe veffels are not of much importance: a fmall degree of re-

fiftance,

fiftance, by a finger, or fome folded li-nen, applied to the wounded part, will ge-nerally ftop the bleeding.

This tranfmiffion of the blood from the heart through the arteries, and back to it by the veins, is the CIRCULATION; which was the difcovery of our illuf-trious countryman, Dr. WILLIAM HAR-VEY.*

It

* The ufe of the lungs in the circulation is here pur-pofely omitted. —— The reader, who is defirous of en-larging his mind with the principal truths of anatomy and phyfiology, will be amply gratified in his inquiries. It is to be lamented that this kind of knowledge is not generally purfued as a part of a liberal education. The ftudy of the animal economy affords the moft beautiful and fatisfactory ideas, and is calculated to prove highly beneficial to fociety; for, it enables men to diftinguifh between ignorance and knowledge, and, confequently, to encourage deferving men, fupprefs quackery, and ad-vance true medical fcience. —— The medical books, that are frequently to be found in the libraries of gentlemen, are likely to produce very different effects. — The fum-mary accounts of difeafes, with receipts for the cure of them, are pillars of the moft dangerous empiricifm: fo

far

It is very plain, then, that, if a bandage or ligature be made sufficiently tight around any limb, the flow of blood into all the parts below will be prevented. But, to render this effect certain, the pressure must be very great in the whole circumference of the limb; and, in some cases, from the situation of arteries between bones, the end cannot be obtained. To perform this process, therefore, successfully, in cases of wounds and operations, and, at the same time, to prevent the evils of an exceedingly strong *general* pressure, surgeons have fixed on certain parts of the TRUNKS of arteries for the application of a pad or COMPRESS. — These parts are expressed in the annexed plate.

The PULSE is the beating, or distending, of an artery, from blood propelled

far from furnishing the mind with useful truths, they fill it with error, and beget a confidence in ignorance often fatal to health and life.

B 3   into

into it by the heart. The fpaces of time between the pulfations are periods when the heart itfelf is diftending with blood returned to it by the veins.

Now it is evident, that there can be no pulfation when the flow of blood and diftention of an artery are prevented. Where, then, a pulfe can conveniently be felt, as in the wrift, the ceafing of it, from a preffure made on the trunk above, will prove that the preffure is made effectually. To illuftrate this by an experiment : — Let a friend feel the pulfe in your wrift; then apply two or three fingers in *the little pit, immediately below the collar-bone, clofe to the fhoulder, marked* a *in the plate.* Prefs ftrongly, and the pulfe will ceafe; becaufe, the artery that fupplies the upper extremity *paffes under the collar-bone, over the firft and fecond ribs, along this part,* and will be now preffed againft one of thefe ribs. Remove the fingers, and again apply them, and

and the pulfe will be found to alternate
with the preffure.

Suppofe, then, a wound to be received,
an artery of a confiderable fize cut or
torn, and a copious bleeding, in confequence, to happen, in any part of the
arm *below* the place *a :* — it is manifeft, that, by making a preffure with
the fingers, in the manner defcribed, or
affifted by a pad between the fingers and
the part, the bleeding would inftantly
ceafe. Is not this an ufeful remark?
Let this little procefs be your firft exercife; and, when you are expert in the
practice of it, we will proceed to confider the other places in the limbs where
effectual compreffion may be made, and
the inftruments proper for the purpofe.

The arteries of the upper extremity or
arm proceed from the trunk at *a,* after
this manner: *the trunk paffes into the arm-
pit, deeply fituated; it then proceeds along
the fide of the arm, next the body, oblique-
ly towards the fore part of the joint or*

B 4   *bend,*

bend, and here divides into three branches. In this courfe to its divifion it lies near the bone, and may therefore be fuccefsfully compreffed. — The fituation of this trunk to its divifion is defcribed in the plate by the lines *b*.

All compreffive means, for preventing a flow of blood from wounded arteries of the upper extremity, muft, therefore, be made either at *a*, or in fome part of the courfe of the trunk of the artery, expreffed by the lines *b, between the arm-pit and the bend of the arm.*

The diftribution of the veffels of the lower extremity is in this way. — The artery paffes from the cavity of the belly to the GROIN, where, in thin perfons, the pulfation of it may be felt.

At this place, in cafe of wound and effufion of blood very high in the thigh, effectual compreffion may be made, by fome fingers preffed very ftrongly, in the manner defcribed for compreffion below the collar-bone ; though it were better

to

to have fome kind of ftrong pad, or firm body, fuch as will be defcribed, interpofed between the fingers and the part.

From the groin, the artery proceeds in an oblique direction, downwards and inwards, as expreffed by the lines c ; *and, at about the middle of the infide of the thigh, expreffed by the comprefs* d, *it lies clofely to the bone.* This is the moft favourable part for making a preffure upon it, becaufe of the refiftance of the thighbone behind. . And, where there are opportunities of choice, as in cafes of wounds or operations *below* this part, this is the place which furgeons fix on for the application of the compreffing body ; it therefore deferves your particular attention.

The courfe of the veffel is then *downwards and backwards to the* HAM ; *in the hollow of which, againft the lower flat part of the thigh-bone,** compreffion

may

---

* It is highly neceffary, that the greateft attention fhould be paid to this point of inftruction. The pad of

the

may again be very fuccefsfully made in all cafes of wounds or operations below the knee-joint. But *beyond* this part compreffion muft not be depended on; for, immediately below the joint, the artery divides, like that of the upper extremity, into three veffels, which are fituated between the bones of the leg.

You have, I doubt not, anticipated me in a remark on the goodnefs of the great CREATOR, in ordaining the fituation of the larger blood-veffels fo that they fhould not be expofed to danger in the neceffary offices of life.

the tourniquet being placed as here directed, the ligature muft be brought *round the thigh, immediately above the knee,* and the twifting, of courfe, be made upon the thigh. If, on the contrary, the pad be placed in the hollow of the joint, and the ligature carried round the leg, the confequence might prove fatal before the error could be corrected. But it is generally more fafe to make compreffion in the middle of the thigh than at the part here defcribed, and more proper as to effects afterwards; for, it is always right that the bruife and irritation that neceffarily arife from the ligature fhould be as diftant as poffible from the feat of injury or operation.

The

The inftrument called TOURNIQUET, we are informed, was the invention of a furgeon, named MORELL, at the fiege of BESANÇON. It confifts of four parts: *viz.* 1. *e*, a yard and half of ftrong worfted, or other kind of band, an inch broad; 2. *f*, a pad of leather, tightly ftuffed with wool or horfe-hair, two or three inches long, and of an inch breadth and thicknefs, having a loop on one fide for the band to be flided through;* 3. *g*, a piece of ftrong leather, three inches long and two broad, having two apertures, an inch afunder, for paffing the band or ligature; 4. *h*, a piece of fmooth, round, and ftrong, wood, about four inches in length.

Defcription often fails even in things of great fimplicity. This may poffibly be the cafe in the account of the TOUR-

* It has been fuggefted, that, for the ufe of perfons who may not retain an accurate remembrance of the fituation of the veffels, it were better for this pad to be made as large again as here defcribed.

NIQUET;

NIQUET; but the flighteft view will make it underftood.* The manner of applying it is this. — Place the pad upon the proper part of the artery to be com-preffed; bring the band, paffed through the loop of the pad, round the limb, and carry the ends through the apertures in the leather; make a double knot with the ends, leaving a fpace between the knot and the leather that will admit three or four fingers; through this fpace pafs the ftick, and with it twift the ligature fufficiently tight to ftop the flow of blood through the artery into the limb. The leather, knot, and twifting, are to be placed and made upon the upper part of the limb, nearly oppofite to the comprefs.

* It is much to be regretted that this inftrument is not generally known, and kept in every family. The price of it is too trifling to be mentioned. — The life of a va-luable gentleman in Hertfordfhire would have been lately loft for want of it, if a furgeon had not providentially called at his feat in the moment of a dreadful effufion of blood, from a wounded artery in his hand, occafioned by the breaking of a bottle in a fall.

It

It is manifeſt that this procefs, ſimple as it is, requires both hands for tying the knot; and, therefore, that you could not apply the tourniquet to your own arm without aſſiſtance. It is as plain, alſo, that it demands a conſtant application of a hand to the ſtick, as the ligature would otherwiſe inſtantly ſlacken.

To obviate the neceſſity of two hands, in regard to the arm, let the ligature be about twelve inches long, and have at each end a loop: proceed in its uſe exactly as already deſcribed; only, inſtead of making a knot over the leather, paſs the ſtick through the loops at the ends of the ligature, and then perform the twiſting.

To fix the ends of the ſtick, ſo as to prevent the ligature from untwiſting, and the conſtant application of a hand, faſten a portion of tape or packthread, by means of a hole, at each end of the ſtick; carry the two pieces round the limb, and ſecure them by tying or pinning. — Many other expedients

expedients may be contrived to anſwer this purpoſe.

Beſides the tourniquet that I have de-ſcribed, there is another, an excellent piece of machinery. It was invented by M. PETIT; and improved by the late Mr. FREKE, of St. Bartholomew's Hoſpital. It need only be ſeen to be underſtood. —— The pad, *i*, being placed upon the artery, and the ligature buckled at *k*, then, by turning the ſcrew, the upper moveable portion, *l*, will be raiſed from the lower, and, conſequently, the ligature may thus be drawn to the degree of tightneſs re-quired.

The advantages of this inſtrument are very great. — It may be applied with only one hand; and, on being fixed, will re-main ſafely in that ſtate without atten-tion.

Thus the defeĉts of the former inſtru-ment are ſupplied; and, on every occa-ſion for a tourniquet, *when there is a want of* ASSISTANTS, nothing more uſeful

was

was ever contrived. The furgeons on-board fhips of war, in the hurry of en-gagement, oftentimes cannot poffibly per-form their neceffary operations fo foon as required : by this machine, the bleeding from wounds can inftantly be reftrained, and then the wounded may wait, without danger, till the furgeons can calmly exe-cute their duty.—Government have wifely directed every fhip to be fupplied with many SCREW-TOURNIQUETS.

And now, young gentlemen, after what has been faid of VESSELS and TOURNI-QUETS, fuppofe any of you were wound-ed by a penknife, or other thing, in the thigh, leg, or arm, and, a large artery be-ing punctured, a violent bleeding fhould enfue. You have no tourniquet ; but you clearly underftand what has been taught on this fubject. How, then, would you act? — Undoubtedly you would inftantly pull off your garter, or take the firft piece of ftring or cord you could find ; roll up your handkerchief hardly, and lay it on the

trunk

trunk of the artery above the wounded part;
paſs the garter or cord over the handkerchief and round the limb ; tie a knot, leaving a proper ſpace ; and then twiſt the ligature by a piece of your ſtick or cane, or
any other firm body you could procure.

It may be truly ſaid, that, in either of
the branches of medicine, " a little learn-
" ing is a dangerous thing." My ſole de-
ſign was, to explain to you the means of
ſtopping a flow of blood from wounded
limbs, and preventing fatal conſequences,
*till more effectual aid from ſurgery be ob-
tained.* It is happy for mankind that there
are profeſſors in this ſcience in almoſt every
town and village, as well as appointed to
the army and navy.

A BRIEF

EXPLANATION

OF

THE NATURE OF WOUNDS,

MORE PARTICULARLY

THOSE RECEIVED FROM FIRE-ARMS.

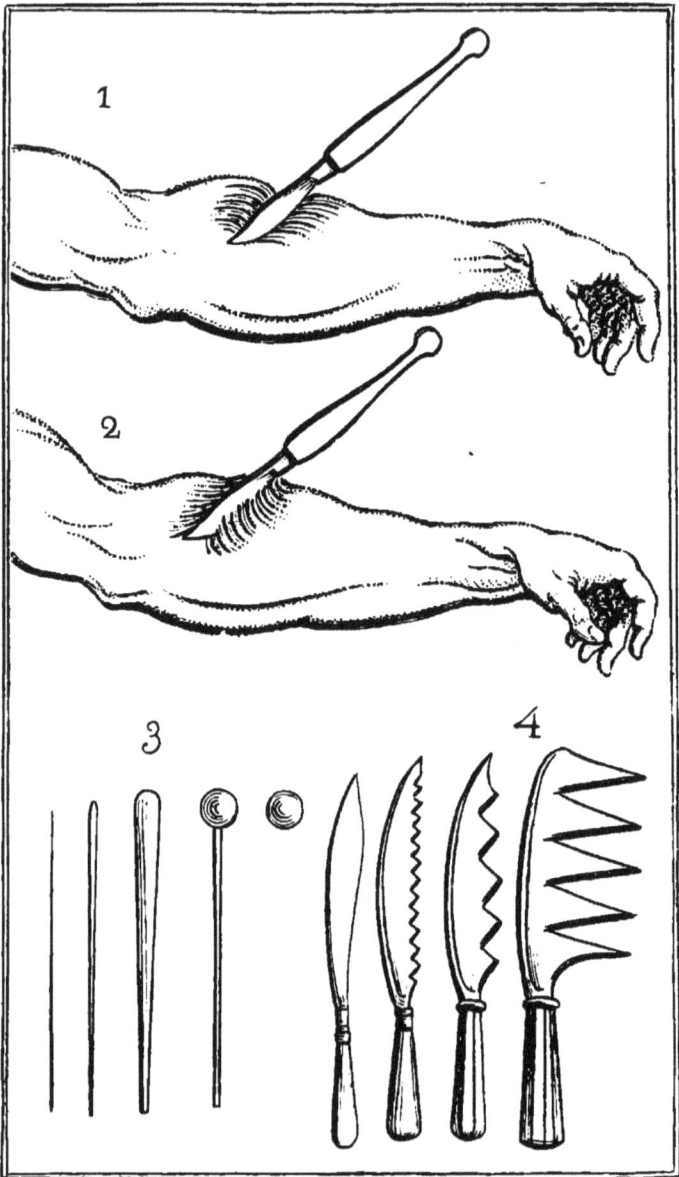

A BRIEF

EXPLANATION

OF

THE NATURE OF WOUNDS.

IT would be fortunate for mankind, if every perſon poſſeſſed that knowledge, by the help of which the intention of a judicious and honeſt ſurgeon might be underſtood and promoted, and the effects of ignorance and impoſition prevented.

A knowledge of the ways in which any part of the body can be divided, leads to

that

that of the nature of wounds ; and this information, added to a very little acquain-
tance with the animal economy, points out
the manner in which fuch fpecies of injury
fhould be treated.

The terms of diftinction applied to
wounds will be more clearly underftood
from confidering the manner in which they
happen.

Conceive, then, the acts of dividing the
fibres of an animal body by an inftrument
moving in a direction either perpendicular
to the furface of the fibres, or parallel to
it.

In the former cafe, the inftrument, of
whatever defcription, muft be preffed perpendicularly to the furface ; from which
preffure, the fibres will be more or lefs
ftretched ; bruifed in a mafs together,
proportionably to the extent of the preffing
body ; and, laftly, broken through in a
perpendicular direction. *(Fig.* 1.)

In the latter cafe, the inftrument muft
have teeth. Thefe teeth muft be made to

enter

enter into fpaces between fome of the fibres ; or they muft prefs down fome fibres,
while others rife into the intervals of the
teeth. The inftrument being then drawn
in a direction parallel to the furface, there
will be a yielding and ftretching of the
fibres, till they can yield and be ftretched no
farther ; and then they will be broken
through in a parallel direction. *(Fig. 2.)*

Hence are derived elementary ideas of
every fort of wounding inftrument, and of
every diftinction of wound.

Suppofe, for illuftration, a feries of inftruments, placed in regular order, beginning with the fineft needle, and ending with
a ftick having a leaden bullet fixed to its
extremity. Then figure to the mind the
fame bullet, unconnected with the ftick or
any other body. Imagine, in the next
place, a wound produced by any one of
thefe inftruments, in a fimilar way, by
preffing, bruifing, and rending, the fibres
of the part, perpendicularly to its furface.
(Fig. 3.)

The

The wound produced by each of thefe inftruments will be a *contufed wound*; but the wound made by the needle will be ftyled a *punctured wound*. Yet this very needle would fatally bruife a minute infect. Whence we learn, that the technical terms *contufed* and *punctured* are relative to the fize of the inftrument, and the tenuity of the part injured.

If the wound made by the bullet fixed to the end of a ftick, by the force of an arm or otherwife, be a contufed wound, it follows, that a wound caufed by the fame bullet, propelled by the force of gun-powder, will alfo be a *contufed wound*, without any difference whatever, fave what may arife from greater force or *momentum*, through greater velocity, and from its feparate ftate, on which account it may be made to penetrate, and may be reflected, differently from what can happen while fixed to a ftick held by a hand.

Suppofe another feries of inftruments, placed in the fame regular order, begin-

ning

ning with the fineft knife, and terminating
with a row of fpikes. Then confider the
analyfis of a knife. It confifts of many
pointed teeth or fpikes; thus it anfwers to
the character of a faw; and, as graduating from the back to the edge, it poffeffes
the principle of a wedge. As the edge is
more or lefs finely graduated, and the points
are more or lefs fmall, fo is the knife expreffed as more or lefs keen. *(Fig. 4.)*

Conceive, now, a wound to be made,
by each of thefe inftruments, upon the
principle of a cutting inftrument. The
points of each muft be preffed down between the fibres, and next drawn in a direction parallel to the furface, as has been
explained.

But how different, in many refpects, will
be the wound made by the fine knife, and
that occafioned by the faw, or inftrument
with fpikes. The wound made with the
leaft conceivable ftretching or bruifing effect, is called a *fimple incifed wound :* as that
made by the hand of a furgeon with a fine

knife ;

knife ; in which cafe, it is gently preffed as. a wedge, fo as to pafs the points of the edge into the fpaces between the fibres to be divided, and no more ; the inftrument is then drawn in a direction parallel to the furface ; the tender fibres are thus broken ; and, by repeated applications and drawings of the knife, fucceffive layers of fibres are divided to the extent required. — The one attended with much violence of effect, from the refiftance of the fibres, as that made with a jagged inftrument, is termed a *lacerated wound*.

But, when we confider the graduation of the fharpeft knife into the fpiked inftrument, and that, in wounding with each as a cutting inftrument, the fame procefs takes place, thefe terms alfo will appear relative, to the finenefs of the inftrument, and the delicacy of. the fubject divided. ———— That which might be expreffed as a *fimple incifed wound*, in the fide of an elephant, would probably be a dreadfully-lacerated one in the human body.

In

In *every* cafe of wound, preffure and ftretching muft happen prior to divifion of fibres, with whatever velocity of fucceffion thefe effects may be produced.

Stretching irritates fibres through their whole extent ; as the ftring of a mufical inftrument is vibrated through its whole length by a force applied to any part of it. Divifion faves from farther ftretching ; as the breaking of a vibrating ftring deftroys the continuity between its extremities, and fets at reft the feparate portions.

The effects of irritation are pain ; inflammation, and its confequences ; convulfion ; delirium ; fpafm ; and locked jaw.

Proportionably to the degree of the irritation ; the irritability of the wounded part, and the body generally ; will be thefe effects, up to death.

Contufion always implies ftretching and irritation, and alfo death of parts. The effects of irritation, and detachment or floughing of dead parts, confequently follow.

The

The procefs by which divided parts are united, and parts loft are fupplied, is univerfally the fame.

The agents are the abforbent-veffels and the arteries. The former labourers being employed in conveying particles away; the latter in bringing and depofiting matter of fupply.

There can be no union of divided parts without a medium of new fubftance. The expreffion, therefore, of a union *fine medio* is founded in error.

A glutinous matter is produced by the extremities of the divided arteries; its properties being, in fome degree, determined by the irritation of the hurt.* Minute veffels, of the three fpecies, (*viz.* arteries, veins, and abforbents,) fhoot into this glu-

---

* So that a wound, hypothetically admitted without irritation, would want the neceffary *ftimulus* to the early fteps of union or fupply. In this remark, however, we have only a particular illuftration of a beneficent general law refpecting the prefervation of every part, and of the whole animal fabric.

ten,

ten, and increafe, till the mafs becomes duly organized for the end required; and the veffels of the fkin have, according to their nature, formed a cuticle or *cicatrix.*

The veffels thus produced fix the ultimate ftate of the new-organized fubftance, according to the difpofition of the veffels they are extended from. Thus, if the divifion be of bone, they will fecrete bony matter, and form a union by what is termed *callus*; and, according to the ftructure and functions of the various other parts of the body, will be the denfity, refiftance, flexibility, &c. of the medium of union, as in mufcle, tendon, ligament, cellular fubftance, membrane, fkin, &c. ; only it muft be obferved, that no part is united or fupplied with a fubftance poffeffing the original characters of the part feparated or divided, excepting cuticle.

In a fimple incifed wound, if the fpace between the divided fibres be very inconfiderable, if there be no extraneous body in

that

that fpace, if irritation and inflammation be not fo great as to produce *pus* or matter, then may the fides unite by what is called the *firſt intention*, or, very properly, *agglutination*; for, they are truly, in the firſt place, glued together.

The objeﬅs of furgery, then, in a fimple incifed wound, are, to reﬅrain irritation and inflammation; to remove extraneous matter; and to bring and retain the fides in conta꜀ﬅ.*

Bleeding,

---

* The lefs the quantity of uniting medium, the lefs liable it will be to change afterwards; the ﬅronger will be the union; and the more perfeﬅ, in every funﬅion, will be the united part. The circulation in a new fubﬅance is never fo ﬅrong as in a part originally formed: whence its veﬅfels are lefs capable of fuﬅaining the influence of caufes produﬅive either of ulceration, by occafioning the abforbents to convey away loaded and oppreffed parts; or of death and floughing, through obﬅruﬅion by preﬅﬅure upon the returning veins and abforbents. — The breaking of the *cicatrices* of wounds and ulcers, the confequence of many caufes affeﬅing the veﬅfels beyond what they can bear, are illuﬅrations of this pofition. — When it is defigned to unite by the firﬅ intention, care fhould

Bleeding, purging, injection, and low regimen, are proper in the firſt intention; adheſive plaſter, bandage, bolſter, ſuture, and, above all, *poſition*, in the laſt.

It muſt be manifeſt, from the nature of what is denominated lacerated wound, that it will be attended with great irritation; the effects of which are, therefore, to be guarded againſt by opium, in addition to the other means mentioned, indicated alſo in this caſe. Fomentation of warm water, bread and milk poultice, or poultice of de-coction of poppy-heads and linſeed-meal, are proper for the purpoſe of allaying irri-tation and pain.

ſhould be taken that the divided ſides are, in every part, brought into accurate contact. It were better that a chaſm ſhould be left near the ſurface, than that the ſuperior parts ſhould be united, while a hollow is left beneath, that will become the ſource of future pain and trouble. Nicety in the application of the edges of the ſkin, however proper with a due regard to the deeper parts, is not, therefore, of ſo much importance as the co-aptation of the ſides from the bottom of the wound.

If

If a mufcle, the fibres of which are united in one tendon, be partially divided, the effects will probably be more violent than if the whole were cut through.

In the cafe of a punctured wound, (by a fmall fword or bayonet for inftance,) no inquiry into its depth or penetration fhould be made, by probing or otherwife. Gratification of curiofity, in this cafe, may prove fatal, but never can be productive of the leaft benefit. Life will often depend entirely upon immediate agglutination ; to promote which, all the means propofed for preventing and removing irritation and inflammation fhould be rigidly employed. A probe would break down the tender glutinous medium, and irritate the fenfible extremities of the divided veffels, upon whofe gentle action fuccefs altogether depends. — There is no cafe in which attention to pofition is more required than in this; and it fhould be remembered that no part about the trunk can be at reft otherwife than in a recumbent fituation.

In

In every contufed wound, there is an object to be regarded, in addition to what oc-
curs in other diftinctions of wounds; namely, the feparation of dead parts. This
procefs being very weakening, reduc-
tion of the ftrength, by bleeding, &c.
fhould not exceed what is abfolutely required on account of an exceffive *degree*
of irritation and inflammation. Soothing
means, as fomentation of warm water, and
poultice with milk or decoction of pop-
py, are generally proper.

In every gun-fhot wound, then, there is
death, and muft be feparation, of parts.

According to the *momentum* of the ball,
and the refiftance it meets with in its pro-
grefs, fo will reflection more or lefs readily
happen; and reflection will, of courfe, be
determined by the angle of incidence.

Perfons, ignorant of the reflections that
are produced upon bullets paffing into or
through any part of the body, have conclu-
ded very falfely concerning the parts injured
in gun-fhot wounds; and, upon the foun-
dation

dation of fuch miftakes, many marvellous ftories are related.

Balls have been reflected round the body, without penetrating the *peritonæum*, or membrane that lines the cavity of the belly, and without perforating the *pleuræ*, or membranes that line the cheft ; and have then either lodged, or paffed out at an oppofite part. The like events have happened refpecting the fcull and its contents. Even a whole charge of flugs, from a blunderbufs, has penetrated one fide of a knee-joint, paffed round the knee and through the oppofite part, without injuring the articulation.

In any fuch like cafe, it is not unufual haftily to conclude, that the ball has gone through the bowels, or the brain : and the laws of the animal economy have been thence mifinterpreted.

It is a vulgar error, that the contents of fire-arms do no harm when difcharged clofely applied to the body.*

When

---

* This opinion was, however, feemingly affented to a few years ago, at the Old Bailey, in the trial of Dr. Elliot

When a bullet penetrates a flexible part, it feldom happens that any portion of the fubftance is detached inwards before the ball; for the divided extremities of the fibres, at the point of rupture, are bent, and yield to the paffing body. The fibres afterwards recover themfelves, according to their degree of elafticity, from their cur-ved ftate, and prefent an aperture bearing but a fmall proportion to the fize of the ball. This is moft remarkable in a mufcular or flefhy part.

When, however, a ball penetrates an inflexible body, as bone, the effect is different: a portion of the fubftance penetrated is forced inwards before the ball. — If a bullet pafs through a hard body, it will fplinter and fcale the furface of its egrefs; while that of its ingrefs prefents an

liot for fhooting at a lady. In confequence of which, experiments were made to afcertain the truth in this matter. The refult was, (as common obfervation and common fenfe led to fuppofe,) the nearer a piftol or gun is applied to any part, when fired, the greater is the effect.

D   opening

opening nearly correfponding with the fize of the bullet. This is illuftrated by the effect of a cannon-ball that has paffed through the fide of a fhip, the fplinters from which are fo dreadfully deftructive. Every cafe of fracture of the fcull affords alfo fome degree of illuftration. Such fracture happens from a force applied to the part itfelf; and the portion of bone beaten inwards, will always be fcaled or fractured farther in the internal than in the external furface.

In every inftance effects will vary, according to circumftances of refiftance in the fubftance penetrated; the figure, and obliquity of direction, of the body penetrating; &c.

In every cafe of gun-fhot wound, whether in a yielding or an unyielding part, extraneous matter may be forced inwards.

There is nothing more myfterious, then, in gun-fhot wounds, than in the other diftinctions of wounds: the *phænomena* of them

them all are explicable by the fame laws of Nature.

There are occafions for the aid of furgery, when advice and affiftance from afar cannot be obtained ; when *inftant* decifion, *immediate* means, are neceffary for preferving life. Of fuch a nature, generally, are wounds from fire-arms. A leffon this to young furgeons who enter into the fervice of the army or navy, in the hope of rifing to fituations of the moft ferious refponfibility.

There are *two* periods, refpecting HÆ-MORRHAGE in gun-fhot wounds, to be particularly regarded. Through the laceration of large blood-veffels, a fatal effufion may *inftantly* happen, if not prevented, at firft, by the tourniquet, or other compreffive means; and next by ligature, operation; &c. —— The fecond period is, when the bruifed and dead parts begin to feparate, or flough away. Openings into veffels, and dangerous hæmorrhage, may thence fuddenly happen ; fo that, in cafes where,

from

from the circumftances of the wound, fuch an event is apprehended, it is to be guarded againft in the moft cautious manner.

There are *two* periods, alfo, in gun-fhot wounds, when endeavours to extract foreign bodies are proper : — Firft, immediately after the accident, before fwelling has taken place ; and, fecondly, when tumefaction, from irritation and inflammation, has fubfided by fuppuration.

Enlargement of the wound, when it can be fafely done, or an opening at a diftant part near which the ball is felt, will often be lefs injurious than repeated introductions of the forceps into the wound made by the ball.

When a ball is lodged out of obfervation, either in the brain or in any part within the belly or cheft, it is not to be blindly fought after. Meafures beft calculated to prevent evils, from the prefence of the extraneous body, are immediately to be adopted. Effects are then to be watched, and made the guide of future conduct.

MR.

Mr. Ranby's book on gun-fhot wounds contains many valuable obfervations, and is written with great fincerity ; but its general doctrines are either unfounded, or not clearly conveyed. The practice of dilating generally, for inftance, is not warranted by reafon or experience ; and the effects of bleeding, and of the bark, are not peculiar and fpecific in this cafe, as might be thought from the tenor of the work ; but are to be accounted for upon thofe known general principles by which fymptoms and their remedies are explained.

Dilatation of the wound fhould be made only when plainly required. It may be neceffary at two periods of time, namely, immediately after the hurt, for the more ready extraction of the ball, or any other thing that may have been forced into the part ; for the more effectual making of ligature, in the cafe of hæmorrhage ; and, fometimes, for the fake of dividing a mufcle entirely:—and after fuppuration, when the fame reafons may call for it as at firft ;

D 3 and

and when, befides, a free exit to matter becomes indifpenfably neceffary. On the laft account, dilatation may frequently be ufeful; for, as the wound is generally zig-zag, through the different degrees of refiftance of parts, matter is very liable to be retained, and, confequently, to require expedients for its difcharge. From the beginning to the end, every caufe of irritation is to be avoided. Operations of any kind are allowable only as far as they promife obvious definable benefit.

Applications to the part fhould be fimple and eafy. Fomentation, and poultice with milk or decoction of poppy-heads, will, at firft, be moft proper; and, when fuppuration is eftablifhed, and the veffels need moderate excitement, a poultice of porter and oatmeal will probably be as good an application as can be employed.

Wounds from fire-arms are practically diftinguifhable into two ftages; the firft terminating, and the fecond commencing, at the period of fuppuration.

During

During the firſt ſtage, the violence of ſymptoms of irritation and inflammation is to be moderated by bleeding, purging, ſmall doſes of antimony, opium, diluting draughts of watery drinks, &c. *Bleeding* ſhould, however, be allowed with the ſtricteſt regard to the pulſe, as expreſſive of the ſtrength of the body. It may be copious at firſt, eſpecially from the divided veſſels themſelves; but it ſhould be repeated rather in moderate quantities than largely. Topical bleeding, by leeches, will prove more immediately beneficial than by the lancet, and leſs weakening in its remote effects.

Inflammation, as neceſſary to ſuppuration, and the detachment of dead parts, *muſt* happen, for the event to be fortunate. If, therefore, the ſtrength be ſo reduced that inflammation cannot be ſuſtained in a due degree, or for a ſufficient length of time, the termination will be fatal.

In every caſe of neceſſarily large detachment of parts, ſuppuration is to be looked

D 4

for

for as an event of the utmoft importance.
Indications will then inftantly change, and
upon anfwering them in time will depend
principally the iffue of the cafe: for, as,
during the firft ftage, means for keeping inflammation within proper bounds are neceffary; fo, when inflammation has terminated in fuppuration; when pain, the
confequence of ftrong action of veffels and
tenfion of parts, and fever, have ceafed;
when thefe fymptoms are fucceeded by a
finking pulfe, general fenfe of weaknefs,
difcharge of matter, and fall of fwelling,
in the feat of injury; the lowering means
are immediately to give place to thofe of
oppofite tendency, — to bark, fmall dofes
of opium, good aliment, fpice, wine, porter, &c.

It, indeed, fometimes happens, from a
previoufly weak ftate, or hæmorrhage from
the wound, that the ftimulants juft mentioned are neceffary from the very beginning. So far from inflammation rifing to
too high a point, it cannot be raifed to, or
retained

retained at, a proper height for all the ends required, through the inflammatory action of the veſſels. This is, indeed, a ſituation of great peril, and calls for the niceſt attention; for, excitement, beyond what the circumſtances of the moment demand, will prove, in effect, a waſte of vital power. The minutes, therefore, muſt be watched ; and according to what they bring forth muſt be determined the adequateneſs of remedies.

AMPUTATION is to be performed only under circumſtances, unequivocally expreſſing it to be neceſſary for the preſervation of life.

Events, in gun-ſhot wounds and compound fractures, ſeem to juſtify the aſſertion, that ſucceſs oftener attends amputation after ſuppuration, than when performed before that period.

There are, however, occurrences in theſe and other deſcriptions of caſes that at once determine the judgement as to the propriety of *immediate* amputation;
and,

and, independently of the hurt, abstract-
edly confidered, there are many things
that will have great weight in deci-
ding upon the operation as the beft ex-
pedient, even when, *prima facie*, the nature of the injury may be fuch as, under more favourable circumftances, might
juftify a lefs fevere decifion. The fituations of wounded people, in a crowded
hofpital, in an airy plain, in the field
of battle, in a chamber of convenience
and fecurity, in the anxious moment of
engagement, when in quiet poffeffion of
the field or the fea, during the hurry
of a purfuit, the alarm of a retreat, &c.
are very different, and will prefent reafons for acting differently in fimilar injuries.

General chirurgical principles, confirm-
ed by experience, muft, however, be ad-
verted to, and fhould be the guide upon
every occafion.

The more topical or limited the hurt,
the more proper, generally, will be *immediate*

mediate amputation; and, *vice versâ*. A wound, by a mu{ket-ball, in the ancle-joint, and one in the thigh, with fracture, from a cannon-ball, are cafes that illuftrate this pofition. — It is the more neceffary that an inexperienced perfon fhould well confider this rule, as the figns of the greater extent and degree of violence might otherwife be very likely to miflead his judgement.

The operation fhould be done completely beyond the feat of *contufion*, as well as of fracture, &c. This plain rule, aifo, is of great importance: the utmoft care, therefore, is neceffary in determining upon the nature and boundary of the injury.

Gun-fhot wounds in the joints generally require amputation.

In every cafe of wound of a large artery, it is fafer to make a ligature upon each divided extremity, than to truft to one only: branches may fupply the low-

er

er portion, and continue or renew hæmorrhage.

The period of feparation of contufed and dead parts muft be religioufly watched. The alarm of bleeding may happen when not expected from any *fign* of contufion ; and life will confequently depend upon immediate affiftance. The retracting of a veffel, or fainting, may fufpend hæmorrhage, that may afterwards occur, and prove fatal.

Whenever LIGATURE can be made in the cafe of an opened artery, it ought to be done. Nothing that bears the title of STYPTIC is to be *depended* upon.

Men fhould be wary how they give their fanction to dependence upon STYPTICS in preference to *certain* means of ftopping hæmorrhage. A little matter will fometimes fuffice to reftrain a bleeding. In an amputation of the leg, below the knee, of a boy eleven years of age, at the London-Hofpital, all the arteries retracted

tracted fo much that not a ligature was
made, and he was foon well. If any
thing called ftyptic had been employed
in this cafe, it would have acquired un-
merited reputation, and the lofs of many
valuable lives might have been the confequence.

Refiftance to a flow of blood may be
made by divers means, that may prove
effectual in bleedings from *fmall* arteries ; but are always to be regarded as
fallacious in divifions of large veffels.

Mealy, and tender fibrous, fubftances,
united with the blood, may form a re-
fifting pafte. Acids, fpirit of wine, &c.
may coagulate the blood, and fo occafion refiftance. Stimulating things may
excite the extremities of divided veffels to
contract, and retract, and thence refiftance may be caufed. Coagulation of
the blood in the coats of a divided artery, as well as in the tube itfelf, and, confequently, death of the veffel, may happen
from heat, and various things called cauftics.

Solutions

Solutions of refins may be decompound-
ed by the blood in the part, and the refinous coagulum may obftruct the divided veffels, as with the compound tincture of gum-benjamin, tincture of myrrh,
&c.; and fome of thefe properties may
be united in the fame article: but experience has demonftrated the fallibility
of all fuch means.

Unhappily, however, there are occafions where ligature cannot be made;
and it *fometimes* happens, that the trial
of a ftyptic may be admiffible, even in
cafes where ligature can be performed.
Oil of turpentine, applied by buttons of
lint, will generally prove the moft effectual article of the clafs of ftyptics: being made hot, its ftyptical property becomes confiderably augmented.

But, moft of all, next to ligature, COM-
PRESSION is to be depended upon. This
may be made by means of compreffes of
linen, lint, &c. either againft the ends of
the veffels, upon their fides, or in both
ways.

ways. *Sponge* is admirably adapted for preffure; but, when it is employed, the *rationale* of its ufe fhould be remembered. The end purpofed will depend upon its elafticity. It is, therefore, to be fo pref-fed into, or upon, the part, as, when expanded, to maintain a proper degree of pref-fure againft the open veffels.

Ligature may be made with the great-eft probability of fuccefs upon any artery of the upper extremity; and upon any artery, below the ham, of the lower extremity; and there is fome probability that ligature may be fuccefsful below the large artery, called *arteria profunda*, that goes off from the artery in the groin: but no perfon is to be fuffered to die by hæmorrhage that can be re-ftrained, from any veffel. What may *pof-fibly* happen cannot be foretold. The very order of things, in the diftribution of the veffels in the part wounded, may be reverfed; and it fhould be remem-bered, that the nourifhment of the parts below

below may be effectuated through the
gradual dilatation of myriads of commu-
nicating fmall veffels, in the ratio of their
diameters, where no large artery, that can
itfelf carry on the circulation, exifts.

The difficulty of effectually fecuring bleeding veffels increafes much by the lofs of time;
efpecially, if irritating ftyptics have been
employed. The adhefions, thickening of
cellular fubftance, &c. that follow, render
it oftentimes no eafy matter to afcertain,
and properly bring to view, the injured
veffel: nor is fuch a ftate fo favourable
for the event of a ligature as the condition
before inflammation. Experience has proved it a fafer general practice, in the cafe
of puncture or divifion of a large blood-
veffel, at once fo far to extend the wound
as to allow of tying the artery with eafe
and proper effect.

The elaftic forceps are convenient for
holding veffels while ligatures are made
upon them; but, in precarious fituations,
it is fafer to ufe the needle and ligature;

ture; taking great care, however, to leave out diftinguifhable nerves. The *many-tailed* flannel bandage is the beft for the thigh after amputation. By cutting off one of the portions of a ligature, the bulk of extraneous matter in a wound is favourably leffened.

There is fometimes a ftate of dreadful apprehenfion, concerning operations, even in perfons of undoubted courage. An occurrence, fome years fince, at the London-Hofpital, will exprefs this in a ftriking manner, and may convey fome inftruction. A foreigner was to have his leg amputated, on account of a difeafe in it through which he was finking. He was fo reduced, that it was determined to perform the operation on his bed. At the moment of proceeding to the incifion, he fuddenly raifed himfelf, fainted, and fell backwards. He thus continued fome minutes, with a pulfe barely perceptible. He then recovered a little, again ftarted up, afked,

E   " Is

" Is my leg off?" and, upon being told
it was not, fainted again. It was judged,
that he would inevitably die without amputation ; that he would probably die
from repeated fainting ; that the operation, performed with due care as to lofs
of blood, would tend to roufe, inftead of
weakening, the action of the heart and veffels ; and that, therefore, it ought to be
performed. It was done, with as much
expedition as poffible. —— The operation was entirely finifhed, when he a-
gain raifed himfelf, and put the quef-
tion as before. Being affured that his
leg was removed, he inftantly became
cheerful, and fainted no more. He left
the hofpital perfectly well ; and always
declared, that he had not the leaft pain-
ful fenfation from, or confcioufnefs of, the
operation.

MORTIFICATION is the death of a portion of the body ; and is the confequence of
any caufe that puts a ftop to the circulation of the blood in it. Thus the ar-

teries of a part, in an over-diftended ftate from inflammation, become incapable of acting upon, and affifting in, the propulfion of their contents ; whence the fides of the veins, and abforbents, are at length fo compreffed as to occafion fuch a degree of obftruction and refiftance in them as the power of the heart is infufficient to furmount : the fluids in the tubes, confequently, become ftagnant. In what manner the bruifing of veffels, coagulation of the juices by heat and other caufes, ligature, &c. occafion mortification, may thence be eafily conceived.

It will, moreover, be plain, that a diminution of power in the heart, and in the veffels of any part, may occafion mortification ; efpecially if, at the fame time, a more than ordinary degree of refiftance is to be overcome.

That mortification fhould often follow gun-fhot wounds, will not, therefore, appear extraordinary.

It

It is trufted, no furgeon need be told,
that amputation will not check mortification : or that the operation is not admiffible till Nature has faid, *Thus far,*
and no farther; till, in fhort, the line
of feparation of dead parts is manifeft. —
There are alfo rules to be obferved equal-
ly interefting as this maxim.

SCARIFICATION cannot be of ufe, except in the dead parts, for the extrication
of putrid air.

The application of oil of turpentine, or
any other powerful ftimulant, while inflammation is prefent, muft prove injurious.

Although high action of the heart and
veffels may, for a while, be concomi-
tant with mortification ; and although,
during fuch a ftate, the object of furgery is to cool and calm the heated fyftem ; yet, fooner or later, the powers
will become depreffed, and require all the
aid of diet and medicine to fuftain them
in their functions. In this ftate, wine,

opium,

opium, and bark, are remedies principally to be depended upon. The two
latter have been experienced of such great
efficacy that some persons have ascribed to
them *specific* virtue ; thus ignorantly ad-
mitting mortification to be a specific difeafe. They are beneficial in this cafe,
upon the fame principles as in debility
from any caufe whatever.

When topical excitement is called for,
oil of turpentine, mixed with olive-oil,
may be proper ; but the natural terebin-
thinate balfams, as balfam of Copaivi, &c.
are generally to be preferred.* Aromatic fomentation, with camphorated spirit sprin-
kled over the flannels ; and poultice of
beer, oatmeal, and pepper in fine pow-
der ; and antifeptics to the dead parts, as
vinegar, diluted mineral acids, spirit of
wine, tincture of myrrh, fermentative articles affording fixed air, &c. will be ufeful.

* *Camphor*, united with gum-arabic and water, in the
form of mucilage, becomes an application very antifep-
tic and kindly ftimulative.

E 3   STIMULANTS,

STIMULANTS, both external and internal, fhould, however, at all times, be nicely graduated according to the effects which they produce. *Debility* in the moving fibres is the confequence of action too long continued, too often repeated, and too ftrongly performed; as well as of oppofite ftates, arifing from torpor, or from defect of exciting power.

In mortification of a limb, when detachment becomes neceffary, care muft be taken that amputation be not deferred till the ftrength is too much exhaufted.

The effects of wounds and contufions from fire-arms are often felt at great diftances of time from the infliction of the injury. The bones are frequently the feat of remote evils from thofe caufes. The *phænomena* of difeafed bone arife from the agency of their arteries and abforbents; or from the death of thefe veffels, their contents, &c. as in *necrofis* or *mortification.* — The analogy of the changes in bone with thofe in the foft parts

is

is exact in every particular. A dead
portion of bone is feparated or *exfoliated*
by becoming firft infulated from the living
parts, through ulceration, by abforbents;
and then pufhed away by an organized
fubftance, the production of arteries. —
Sometimes, all the particles of a dead portion are removed by abforbents; and
the event of exfoliation is thence obviated.
Nature often extends the growth of an organized fubftance very far, in order to apply her ufeful inftruments, the abforbents,
to perform important offices; and, not un-
often, her beneficent intentions are fruftra-
ted, in the deftruction of this fubftance,
by ignorance, under the title of *fungus*.
The period of the infulated or exfoliated
ftate of a dead portion of bone fhould be
timely afcertained: then it is that furgery
may be eminently ufeful; for, fuch portion will be felt as an extraneous body,
and upon its fpeedy removal the fate of a
limb, or even life itfelf, may depend.

E 4 When

When a ball, or any other thing, becomes lodged, it often happens, that the veffels of the furrounding parts, when the furprife from its introduction is over, fet about accommodating themfelves to the prefence of the body. The confequence of their friendly exertion is, a callofity of the immediately-furrounding furface, graduating to the naturally flexible parts. Thus the ball obtains a bed, that fhields the more diftant and tender parts from the effects of its preffure. Bodies are more likely to remain quiet in cellular and fatty parts than when near mufcle. — A hollow, with fuch a callous ftate of the fides, arifing from the preffure of retained matter, is termed *fiftula*.

Removal of the thing that preffes, and prevention of future preffure, are all that are required towards relief.

Inflammation in the callous parts, from adventitious caufes, during the refidence of a ball, &c. may produce a train of ferious evils, that can be prevent-

ed

ed or remedied only by the extraction or difcharge of the body.

Injuries of large nerves, particularly in the extremities, and more efpecially in the fingers and toes, are moft likely to occafion locked jaw: injuries of mufcle may be ranked in the next degree among the caufes of this dreadful fymptom.

That ftate of debility, joined with irritability, produced by warmth of climate, particularly difpofes to fpafm and convulfion; which fhould be guarded againft accordingly.

To obviate thefe evils, the fyftem is to be fortified by a generous regimen, bark, &c. while every thing befides is done that is calculated to allay irritation.

All the parts about the head are highly organized, and thence liable to ftrong inflammation when injured. Evacuations are, therefore, to be early and copious.

Hurts received in the head, as well as the trunk, are dangerous, alfo, from the proximity of parts important to life.

BURNS

BURNS from gun-powder are to be regarded in the fame light as burns from
any other caufe. The *degree* of injury
from HEAT, by whatever *medium* applied, is the principal point to be confidered. —— In every cafe, there is irritation: this may be with or without ve-
fication ; and with or without coagulation
of the juices, or death of fome portion.
The heat may alfo be fo intenfe, as at
once to decompofe and crifp up the parts
to which it is applied.

In flight burns, cold fpring water and
vinegar, or cold water alone, applied by
means of linen kept conftantly moift,
generally prove good remedies. In burns
of greater degree, it may alfo be ufeful during the firft two or three days ; when it
fhould he fucceeded by fomentation and
poultice.

This cafe, alfo, is diftinguifhable into two
ftages. All that is appropriate for leffen-
ing irritation is to be done in the firft
period ; and the commencement of fuppuration,

puration, and floughing, may be confidered as the index of the fecond ftage, when wine, bark, &c. will probably be necef-fary.

IRRITATION in the higheft degree, as relative to the irritable ftate of the body, occafions fpeedy death. The nature of the irritant makes little difference. A fcald, in a flight degree, through its extenfive-nefs over the fenfible fkin, may quick-ly caufe death. The abolition of the vital functions, from this caufe, from arfenic, corrofive fublimate, &c. when happening foon from mere irritation, is to be accounted for upon the fame princi-ple.

Wounds, of every defcription, and in-juries from heat, lead to various ftates that require the niceft chirurgical regard. The condition of PURULENT SORES, for in-ftance, is what calls for conftant attention. Some obfervations upon that fubject may not, therefore, be unacceptable.

The

The operations of nature are hidden from human fight, although her agents be fometimes known by their effects. In the forming and moulding of parts in animals, the arteries and abforbents are understood to be the immediate inftruments; but this is learned from the works of thefe veffels, and not from obfervations upon their actions. The organizing procefs of uniting parts, and fupplying deficiency, is veiled by a covering of matter. This may be confidered as a matrix into which arteries, veins, and abforbents, are extended: its condition is, therefore, a point of important confideration; for, it expreffes the difpofition of the veffels that are to organize. Thefe veffels may be influenced by internal means, circuitoufly applied; and, alfo, more directly, by external applications. To determine with judgement upon the latter, it is neceffary to have correct general ideas of the various ways in which things, when applied, act upon the

veffels

veſſels that firſt ſecrete pus, and afterwards ramify into this fluid.

Things applied to purulent ſurfaces are of a nature miſcible with pus, becoming conſtituents of it, as watery, ſpirituous, and ſaline, preparations; or immiſcible with it, not altering its properties, only retaining it, and defending the tender ſurface, as oily and waxen ſubſtances. They may conſiſt of parts, ſome of them miſcible, others immiſcible, with pus; the latter, by warmth and confinement, ſeparating from the former, as in ointments containing metallic and other ſalts. Subtle, oily, and aërial, particles may be diſentangled from ſubſtances in which they were involved when applied, and may either unite with, or penetrate through, the medium of pus, and ſo, or in both ways, act upon the veſſels of the ſurface; as when the eſſential oil of turpentine is evolved from reſinous articles, or fixed air is extricated from fermenting ſubſtances. Alſo, ſome articles act upon ſores in a

manner

manner purely mechanical, as lint, linen, filk, fponge, &c.

From what has been remarked, it will appear, that the pus prepared by the veffels of the part itfelf is at all times the *immediate* and proper covering and defence of the fore furface ; and, confequently, as there is this medium, that chirurgical dreffings do not come into contact with the granulated furface, otherwife than in the manner explained, either by uniting with, or penetrating through, the matter ; except at the moment of applying them, after it has been wiped or wafhed away.

It will be alfo plain, that to bring about, and to maintain, a proper fecretion of pus, are the fimple objects of furgery in the treatment of purulent fores ; becaufe, healing proceeds properly, while the fecreted matter is prepared in due quality and quantity.

Gentle ftimulants are often of great utility in keeping up the healing procefs. The lunar ftone, *(argentum nitratum,)* applied,

applied, in the lighteft manner, to the furface of the fore, avoiding the edges, is
particularly beneficial. Other ftimulants
will, however, fometimes prove more ufeful in changing the difpofition of the veffels.

The effect of acids, in correcting the
air when abounding with putrid effluvia,
particularly in their concentrated ftate, and
elevated in vapour, is well known ; but
the favourable influence of *vinegar* in
SORES is not generally underftood. The
practice, as it has been many years fol-
lowed at the London-Hofpital, confifts in
the application of linen, frequently wet-
ted with a mixture of one part of common
vinegar and two parts of frefh fpring
water. The frefhnefs of the water adds
much to the efficacy of the remedy :
no more, therefore, fhould be mixed
than is required for immediate ufe, as
the water fhould be inftantly drawn from
the well. Diftilled vinegar, and river or
rain water, may, however, prove ufeful
fubftitutes.

fubftitutes. In the fummer-feafon, in hot climates, when putrefcency is to be counteracted locally, or in the furrounding air ; or when a cuticle only is required upon an organized furface, this topical mode of treatment will deferve attention. No perfon of underftanding, however, would apply even vinegar and water to a fore without due advice, provided it could be obtained : for as all effects are relative to circumftances, much judgement is often required in determining upon the fafety of the moft fimple means.

Vegetable applications are frequently preferable to thofe of an unctuous nature. Some plants afford dreffings in their foliage entire, as the cabbage, mallow, plantain, &c. ; the leaves of others are applicable when bruifed, as thofe of hemlock ; and many roots, frefh or boiled, pounded, &c. have proved efficacious in mending the condition of a purulent furface, as carrot, potatoe, onion, &c.

CONTUSION,

CONTUSION, from whatever caufe, dif-
fers only in degree. It frequently happens
from fpent balls, fragments of fhells, and
fplinters. The veffels of the contufed part
may be merely irritated ; may have blood
forced, through their open extremities
or ruptured fides, into the cellular fubftance; or may be diforganized, and de-
ftroyed as living tubes. In each cafe,
there muft be irritation ; to allay which,
bleeding, purging, and opium, are neceffary: and, in the view of promoting
the abforption of extravafated blood, thefe
evacuations are principally to be depended
upon. The beft immediate applications
are thofe endued with aftringency: vinegar, water, and fpirit, are proper. Afterwards, fomentation, poultice, embroca-
tions, &c. may be neceffary. Collections
of extravafated blood fhould not be opened
without abfolute neceffity. The utmoft
exertion fhould be made to obtain the
removal of the fluid by abforption. A
wound, made to difcharge blood from a

F

bruifed

bruifed part, generally becomes ill-condi-
tioned, and fometimes proves fatal. When
the violence of contufion is fuch as to de-
ftroy parts, the period of their feparation
is to be looked to with a watchful eye; for,
evacuations beyond that time, and in the
interval of it, farther than fymptoms abfolutely demand, would be dangerous. The
ftate of parts fatally contufed, yet remain-
ing entire, has fometimes deceived unwary
obfervers.

It may be ufeful to add a word of
caution refpecting the ufe of *vulnerary balfam, tincture of myrrh,* &c. in fimple
wounds. Not a drop is to be fuffered
to pafs into the wound; for, it would
irritate, form a medium of varnifh over
the furfaces of the divided parts, and
fruftrate the intention of its application.
The fides and lips of the wound are to be
accurately clofed, and retained in that
ftate; when fome lint, over the line of
contact, is to be moiftened with the refinous
folution, and fuffered to dry and harden. In
this

this manner, the parts will be defended from the air, and kept in a ſtate favourable for healing.

"I cannot cloſe this tract without obſerving, with much ſatisfaction, that the attention of late paid to the ſurgeons in the army and navy is founded in juſtice to thoſe perſons, and will conduce greatly to the public benefit; for, however ſome men may be irreſiſtibly impelled to ſtudies by their attachment to the objects of them, it generally happens that knowledge is purſued according to the eſtimation in which it is held by thoſe who have the power of aſſigning rank and reward to its poſſeſſors. *Examinations* are proper teſts of that degree of talent *below* which none ſhould be admitted to ſituations upon which the health, happineſs, and lives, of men depend: but examinations can neither create abilities nor direct their application where moſt required. Proper encouragement will do both: it will ſtimulate pupils to apply to the ſtudy of ſurgery with induſtry,

duftry, fpirit, and effect ; and, when maf-
ters in the fcience, it will induce them to
employ their fkill in the comfort and prefervation of thofe members of the commu-
nity who are entitled to our firft and great-
eft care.

THE END.